PORTRAITS
for Classroom Bulletin Boards

Mathematicians • Book 2

Susan and John Edeen

DALE SEYMOUR PUBLICATIONS

Editorial assistance and research: Karen Edmonds
Cover design: John Edeen

Order number DS11901
ISBN 0-86651-456-2

5 6 7 8 9 10 11 12 13-MG-95 94 93 92 91

Introduction

Back-to-School Night is coming and you need a bulletin board idea. That's a perfect time to pull out *Portraits for Classroom Bulletin Boards*. Simply photocopy any or all of the fifteen portraits and biographies in this book—mathematicians whose special talents and achievements have kept their names alive over the centuries.

Be creative to make an attractive display. You can create dramatic effects by copying the pages on brightly colored paper. Or, use white paper and mount the sheets on backgrounds of different colors.

Besides creating something that administrators, other teachers, and parents will admire, you can use the portraits to inspire your students as well. Display them throughout the year to bring to life these famous names from the pages of mathematical history.

You can even turn these portraits into a research-writing activity. The biographies provided here are necessarily very limited; most suggest topics that could be the subject of further inquiry. Challenge your class to see what they can discover in the library about each of these mathematicians and their life's work. Ask students to write their own paragraphs—then display their work alongside the appropriate portraits on your bulletin board.

You might also put together books to display your classes' research work. Give students copies of the smaller portraits found at the back of this book. These can be colored with markers and used as illustrations for the student reports. Collect the reports in binders for display at your open house. Teachers and parents alike will enjoy flipping through the books made by your different classes; students, too, will be interested in seeing each other's work.

With *Portraits for Classroom Bulletin Boards,* you get your students involved in creating graphically appealing bulletin boards. At the same time, you improve their cultural literacy by acquainting them with legendary names from the history of Western civilization.

Portraits are also available for scientists, authors, poets, and artists.

Mathematicians • Book 2

Short biographies and portraits for the following mathematicians appear in this book in chronological order.

Johannes Kepler (1571–1630)

René Descartes (1596–1650)

Bonaventura Cavalieri (1598–1647)

Pierre de Fermat (1601–1665)

John Wallis (1616–1703)

Blaise Pascal (1623–1662)

James Gregory (1638–1675)

Sir Isaac Newton (1642–1727)

Gottfried Leibniz (1646–1716)

Marquis de L'Hospital (1661–1704)

Johann Bernoulli (1667–1748)

Leonhard Euler (1707–1783)

Joesph Lagrange (1736–1813)

Pierre de Laplace (1749–1827)

Carl Gauss (1777–1855)

Johannes Kepler (1571–1630) was a German mathematician and astronomer. As a student, he learned the principles of the Copernican system, which he admired for its mathematical simplicity. After five years as a professor at the University of Graz, Kepler joined the famous astronomer Tycho Brahe at the observatory in Prague. When Brahe died, Kepler took over his position as imperial mathematician and court astronomer.

With Brahe's death, Kepler inherited not only his position, but also his large and accurate collection of data on the motion of planets. Deeply interested in the mathematics of how planets move in their orbits, Kepler went on to formulate and prove the three important laws of planetary motion.

Like Plato and Pythagoras before him, Kepler felt that the world was created according to an underlying mathematical harmony. He combined this feeling with a great capacity for careful analysis and organization of data, bringing a scientific approach to the study of the stars.

Although famous for his work in astronomy, Kepler is also known for developing a system that was a forerunner of calculus.

Johannes Kepler

René Descartes (1596–1650), the French philosopher, developed an early interest in mathematics that remained with him throughout his life. Having delicate health as a child, he was allowed to lie in bed until late in the morning; some years later he remarked to Pascal that the only way to do good work in mathematics and preserve one's health was never to get up in the morning before one felt inclined to do so.

Even as a boy, Descartes liked mathematics because of the certainty of its proofs and the evidence of its reasoning. Insisting on clear and logical thinking, Descartes believed that all experience is organized according to laws that are basically mathematical. So diverse was his intellectual capacity that philosophers, physicists, and mathematicians all look upon Descartes as one of the greatest geniuses in their field. And each group is correct—his brilliance spanned many subjects.

His discovery of analytic geometry—a means of applying algebra to geometric problems—was hailed as one of the most remarkable feats in mathematical history.

René Descartes

Bonaventura Cavalieri (1598–1647), an Italian mathematician, was a pupil of Galileo. A professor of mathematics at the University of Bologna from 1629 to the time of his death, he was largely responsible for the early introduction of logarithms into Italy.

Cavalieri was the most influential Italian writer of mathematical science in the 17th century, writing works on conics, trigonometry, optics, astronomy, and astrology. His greatest contribution was a treatise, published in its first form in 1635, devoted to his principle of indivisibles. This theory treated a line as made up of an infinite number of points; a plane, an infinite number of lines; and a solid, an infinite number of planes. His method represents a crude form of calculus.

Cavalieri's work in this area, while superseded by new developments at the beginning of the 18th century, was nonetheless one of the first steps toward the formation of the calculus. Newton and Leibniz are credited with later inventing the calculus, but the work of both Cavalieri and Kepler pioneered the way.

Bonaventura Cavalieri

Pierre de Fermat (1601–1665) was not a professional mathematician, but rather a lawyer and royal councilor at the parliament of Toulouse, France. However, he devoted his leisure time to mathematics and, even though he regarded it as a recreation, did much original work in the field.

Fermat's interest in the theory of numbers led to several discoveries in this area. In fact, Fermat is considered the greatest writer on the theory of numbers since the time of ancient Greece, and is known as the Father of Modern Number Theory.

Together with his friend Pascal, Fermat is credited with originating the theory of probability, which is used today in such applications as statistics and the insurance industry. He is also said to have discovered the fundamental principles of analytic geometry independently of Descartes.

A prolific writer of more than 3000 papers and notes, Fermat was a quiet and retiring man who published practically none of his work during his lifetime. He did, however, correspond with many of the leading mathematicians of his day, and in that way both influenced and was influenced by his contemporaries.

Pierre de Fermat

John Wallis (1616–1703) was just one of many illustrious men in the 17th century who advanced the field of mathematics with new discoveries. Wallis studied theology at Cambridge and was later ordained and named chaplain to the king. His main interests, however, were in mathematics, and in 1649 he became professor of geometry at Oxford, a position he held until his death.

Wallis was the first to recognize the significance of negative exponents (x^{-1}) and fractional exponents ($x^{1/2}$). He introduced infinite series and was the first to use the symbol for infinity (∞). His main contribution was to systematize and extend the methods of analysis devised by Cavalieri and Descartes.

A voluminous writer with interests in many diverse areas, Wallis wrote on theology, logic, grammar, astronomy, botany, mechanics, the tides, music, the calendar, geology, and the compass. He also devised a system of teaching deaf-mutes. A clever cryptologist, Wallis helped his government decipher coded diplomatic messages. His *Treatise of Algebra* was the first attempt in England to compile a history of mathematics; it constituted a valuable resource of historical information.

John Wallis

Blaise Pascal (1623–1662) was a French scientist, philosopher, and mathematician. As a child he was tutored by his father, who first avoided teaching mathematics to Pascal for fear that it might overstrain his young mind. But Pascal was curious to know what geometry was all about and taught it to himself. He proved to be a math prodigy, and at the age of 16 discovered one of the basic theorems of projective geometry. At 18, he invented the first mechanical adding machine.

Working with his friend Fermat, Pascal formulated the modern theory of probability and combinatorial analysis. Their probability theory arose from an attempt to deal mathematically with a gambling problem. While working on this, Pascal investigated and wrote extensively on a number system that we now call Pascal's triangle.

Pascal's triangle is a triangular array of numbers that dates back to ancient China. Over the centuries, mathematicians have discovered in this triangle many different number patterns related to situations in arithmetic, algebra, geometry, and other branches of mathematics. It is frequently used to determine probabilities and to solve problems involving permutations and combinations.

Blaise Pascal

James Gregory (1638–1675) was one of the first Scots to make a name for himself in both mathematics and physics. After living for some years in Italy, where he studied at the University of Padua, Gregory returned to Scotland to become professor of mathematics at the St. Andrews school and later at the University of Edinburgh.

Much of Gregory's work in mathematics related to the circle. He established a theorem that many others have used to calculate the numerical value of π (pi). His work in pure mathematics showed great originality, and it has been suggested that if he had lived longer, he might have shared the invention of the calculus with Newton and Leibniz.

As it is, Gregory is most renowned for his work in astronomy—in particular, for his invention of the reflecting telescope that bears his name (the Gregorian telescope). He did not actually construct this telescope, but described it fully in his *Optica promota.* Gregory also originated a method of estimating the distances of stars. His astronomical work caused him to suffer severe eyestrain, and he went blind shortly before he died at the young age of 37.

James Gregory

Sir Isaac Newton (1642–1727) was an English mathematician, physicist, and astronomer. His family intended that he devote his life to farming; instead, he became probably the most brilliant mathematician the world has yet produced.

Newton introduced his laws of gravitation in *Principia,* which is considered the most influential and most admired work in the history of science. His other great accomplishment was the invention of integral and differential calculus.

Hating controversy, Newton was reluctant to publish any of his findings; almost all his ideas remained unpublished for many years. This caused problems, the greatest of which was a dispute with Leibniz over which man was the true originator of calculus. Newton seemed to have been working on it first, but he published his findings after Leibniz. Apparently the two men created their systems independently of one another.

Despite such arguments, Newton was a modest man, generous to those who helped him in his work. He is credited with saying, "If I have seen a little farther than others, it is because I have stood on the shoulders of giants."

Isaac Newton

Gottfried Leibniz (1646–1716) was the only notable mathematician produced by Germany in the 17th century. He was also a philosopher, known for his belief that this is "the best of all possible worlds," an idea that was later satirized by Voltaire in his novel, *Candide.*

The most outstanding contribution Leibniz made in mathematics was his discovery in 1675 of the fundamental principles of infinitesimal calculus. Newton had invented his own system of calculus in 1666, but Leibniz published his findings nine years before Newton's system appeared in print. The two systems differed in notation, but some charged that one man had derived the basic idea from the other and just invented another notation for it. Today it is recognized that the two men made independent discoveries. Leibniz's system proved superior to Newton's and was later adopted universally.

Among his other accomplishments was the invention of a machine capable of multiplying, dividing, and calculating square roots. Leibniz also introduced and developed much of the notation we use today, including the decimal point, the equals sign, the colon to indicate division and ratio, and the single dot to indicate multiplication.

Gottfried Leibniz

Marquis de L'Hospital (1661–1704), born in Paris, became interested in the study of mathematics at an early age. He was just 15 when he overheard some mathematicians discussing a difficult problem of Pascal's. Much to their surprise, the young L'Hospital said that he could solve the problem—and then proceeded to do so within just a few days.

When L'Hospital's poor eyesight prevented him from pursuing the military career he had planned, the young man turned to the study of mathematics. He was one of the first people to study calculus as a student of the Swiss mathematician Johann Bernoulli. At the time, there was no textbook on the subject. In fact only a handful of people—Johann and Jacob Bernoulli, Newton, and Leibniz— had any working knowledge of this new branch of mathematics.

L'Hospital became famous in 1696 when he wrote the first textbook on differential calculus, *Analyse des infiniment petits*. It remained the standard work on the subject for most of the 18th century, not only in France but throughout much of Europe as well.

Marquis de L'Hospital

Johann Bernoulli (1667–1748) was one of eight Swiss mathematicians produced by a remarkable family over the course of three generations. A professor of mathematics at Groningen and later at the University of Basel, Bernoulli was one of the most successful teachers of his time, able to arouse in his students the same passionate love of mathematics that he himself felt.

Johann shared in the discoveries of his older brother Jacob, but wrote on an even wider range of mathematical topics himself. Even though the two were at times bitter rivals, for Johann was jealous and resentful of Jacob's mathematical brilliance, they maintained an almost constant exchange of ideas with one another, as well as with such contemporaries as Leibniz.

Bernoulli was among the first mathematicians to realize the surprising power of calculus and to apply it as a tool to a great diversity of problems. His contributions greatly enriched calculus, and Bernoulli helped make its power appreciated in Europe.

Johann Bernoulli

Leonhard Euler (1707–1783), born in Switzerland, was without a doubt the most productive mathematician of the 18th century. He studied under Johann Bernoulli at the University of Basel, obtaining a masters degree at age 16. Euler followed Bernoulli's two sons to Russia, where he taught mathematics and physics at the St. Petersburg Academy.

In a lifetime devoted to work in both pure and applied mathematics, Euler wrote more than 800 books and papers. He made contributions in virtually every branch of mathematics that was then known. In his popular and prestigious textbooks, Euler succeeded in clarifying, expanding, summing up, and ordering all the work of his predecessors.

Euler's writings included the first text on analytic geometry. His work on trigonometry offered concepts and a notation that we still use today. In algebra, several equations bear his name; he also made discoveries related to the theory of numbers; and his work on differential equations is the model for present-day elementary texts. After losing his sight in one eye at age 25, Euler became totally blind later in life but still continued writing, dictating ideas from his remarkable memory.

Leonhard Euler

Joseph Lagrange (1736–1813), the 18th century's greatest mathematician, was a Frenchman born in Italy. He showed no interest in mathematics until he was 17, but then began to study it with such zeal that in just nine years he became the greatest living scholar in the field. Unfortunately, this intense work overloaded his nervous system, leaving him frail and subject to bouts of depression.

Lagrange was teaching in Italy when Frederick the Great of Prussia wrote to him that "the greatest king in Europe" wanted "the greatest mathematician in Europe" to join his court. Lagrange accepted the offer and spent 20 years in Berlin, succeeding Euler at the Berlin Academy. Later, as president of the commission for the reform of weights and measures in Paris, Lagrange was instrumental in establishing the decimal system of metric units.

A student of pure mathematics, Lagrange laid out the principles of a new calculus, the calculus of variations. In his greatest work, *Mécanique analytique,* his treatment of dynamics overturned some of Newton's theories of 100 years earlier. Lagrange also made theoretical calculations on the motions of the moon and planets, and did research on vibrating strings and the nature of sound.

Joseph Lagrange

Pierre de Laplace (1749–1827) was a French mathematician whose major contributions were in astronomy. Born a poor farm boy, Laplace was educated by wealthy benefactors who were impressed by his intellectual promise. He went on to teach mathematics at the military school in Paris.

Laplace studied celestial mechanics with great success; his five-volume *Mécanique céleste* earned him the title "the Newton of France." Laplace developed a mathematical analysis of the system of gravitational astronomy worked out by Newton, explaining the movement of the moon about the earth and the orbital motions of the planets, and demonstrating the stability of the solar system.

Scholars have found Laplace's work difficult to follow. When sure of his end results, Laplace had no patience for explaining the steps that took him there. In his writing, he often covers himself with the phrase, "It can be plainly seen . . . " when in fact there are many links missing in his chain of reasoning. Even so, his research—building on the work of Lagrange and earlier mathematicians—firmly established Newton's gravitational theory and earned Laplace recognition as the unrivaled master of the subject.

Pierre de Laplace

Carl Gauss (1777–1855), a German physicist, mathematician, and astronomer, is ranked with Archimedes and Newton as one of the greatest mathematicians of all time. Gauss began making important discoveries when he was just 17. For his doctoral thesis he submitted the first rigorous proof of "the fundamental theorem of algebra."

Fascinated by number theory, Gauss once stated, "Mathematics is the queen of the sciences, and the theory of numbers is the queen of mathematics." His work in this area is considered the beginning of modern number theory.

In probability theory, Gauss developed the method of least squares and the fundamental laws of probability distribution. Unfortunately Gauss was reluctant to publish any of his findings without complete proof, so many of his discoveries were never credited to him, including his innovative work in developing a non-Euclidean geometry.

Gauss was the last of the great scholars whose interests spanned virtually every branch of mathematics; since his time, the branches have expanded so greatly that students of mathematics generally choose to specialize in a particular area.

Carl Gauss

Johannes Kepler

René Descartes

Bonaventura Cavalieri

Pierre de Fermat

John Wallis

Blaise Pascal

James Gregory

Isaac Newton

Gottfried Leibniz

Marquis de L'Hospital

Johann Bernoulli

Leonhard Euler

Joseph Lagrange

Pierre de Laplace

Carl Gauss